Weather in the Southwest

by Jim Woodmencey

Southwest Parks and Monuments Association

Tucson, Arizona

Bean

Altostratus clouds, which form about 6,500 feet above the earth's surface, above Utah's Sevier Plateau

A winter storm closes in on the southern Arizona grasslands

The southwestern United States is renowned for its great weather, with abundant sunshine and low humidity. For the majority of the year, the Southwest enjoys temperatures warmer than anywhere else in the country, with a propensity for continual blue-sky days, accented by evenings of magnificent sunsets.

This weather creates a rather dry environment, where evaporation outpaces precipitation on an annual basis. But that is not to say that the Southwest is comprised solely of searing heat and parched surfaces. The landscape—and the weather—are far more diverse.

Traveling from the desert floor through the uplands and into the mountains, you find towering red rock formations, mesas of piñon and juniper trees, and snowcapped peaks. You ascend through several distinct climates, experiencing all manner of weather.

Mother Nature shows her fury in this usually tranquil place. Awesome lightning-filled thunderstorms can develop, powerful flash floods rapidly rise, an occasional dust storm may stir, and often snow falls in this part of the country.

As you may discover if you have the opportunity to stay here long enough, the weather in the Southwest reveals itself in ways as varied and spectacular as the land itself.

Climate

At the Grand Canyon the multi-colored layers of rock that make up the canyon walls display the history of the Southwest's climate through geology. These layers of sandstone and limestone tell of a time when a great ocean covered this land, a time when dinosaurs roamed here, and a time when it was much wetter and the vegetation much more lush than it is today.

Very slow changes in the composition of our atmosphere and subtle changes in the tilt of the earth and in the earth's orbital path around the sun account for the periodic fluctuations in our planet's climate. This has led us to the generally warm and dry climate we experience in the Southwest today.

The history of our present climate is short, in the geologic sense of time, not much more than 100 years of official weather records for most locations throughout the Southwest. These day-to-day records, the highs and lows, the droughts and deluges, define today's southwestern climatic record.

The rocky layers of Grand Canyon National Park, Arizona, reveal hundreds of thousands of years of climatic change.

Arid and Semi-Arid

Most of the desert environment in the southwestern United States lies below 5,000 feet and can be classified as either "arid" or "semi-arid." The arid regions are the true deserts, and they are found at elevations below 3,500 feet. This includes the Mojave Desert, the Sonoran Desert, and parts of the Chihuahuan Desert. Most of the year, temperatures are warm and rarely dip below freezing.

The semi-arid regions include the uplands or high desert country, at elevations typically between 3,500 and 7,500 feet. They encompass areas such as the Colorado Plateau, most of the Four Corners region, and western and central New Mexico. Average daytime high temperatures are generally cooler than in the low deserts, and below-freezing temperatures are more common, especially in the winter months when snow may fall.

Aridity does not mean a lack of vegetation. Mexican gold poppies carpet Organ Pipe Cactus National Monument, Arizona.

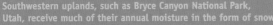
Southwestern uplands, such as Bryce Canyon National Park, Utah, receive much of their annual moisture in the form of snow.

Climate is day-to-day weather averaged over time.

Arid climates receive fewer than 10 inches of annual rainfall.

Semi-arid climates receive between 10 and 20 inches of annual rainfall.

Dykinga

Santa Catalina Mountains, Finger Rock Canyon, Arizona

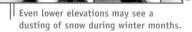

Huey

Even lower elevations may see a
dusting of snow during winter months.

Uplands and the Alpine Zone

Abundant mountains and mesas surround most of the deserts in the Southwest, and the transition from cactus and sagebrush up to wildflowers and pine trees can be dramatic. Within these various mountain ranges and mesa tops, at elevations between 7,000 and 9,000 feet, annual precipitation increases three-fold from the valleys below. Temperatures are much cooler in these uplands, especially in the alpine zone.

Mt. Lemmon, in the Santa Catalina Mountains north of Tucson, is the southernmost ski area in the United States. Mt. Lemmon's annual average precipitation is 35 inches. Tucson's annual average is 12 inches.

During the winter months, snow will accumulate at elevations above 5,000 feet. At even higher elevations, snowfall is vital to both the forests that grow there and to the drier valleys below where it trickles down as snowmelt. Below-freezing overnight lows are likely throughout the winter season and possible every month of the year.

Latitude and Altitude

The climate changes as you change latitude or altitude. As you travel north from the equator (or south for that matter) the sun sits lower in the sky at noon and the overall amount of daily sunlight decreases. Most of the southwestern United States is located in the mid-latitudes, between twenty-nine and thirty-nine degrees north latitude, and receives much more sun—thus warmer temperatures—than the more northern latitudes.

Elevations in the Southwest range from 282 feet below sea level in Death Valley National Park up through 4,000 to 8,000 feet on the Colorado Plateau and reach a maximum of more than 14,000 feet above sea level in the San Juan Mountains.

Increases in elevation change climate dramatically, providing different habitats for a wide variety of plants and animals as you climb from desert floor to mountaintop. Paloverdes and saguaro cactus, roadrunners and jackrabbits are all unique to the lower deserts of the Southwest. As you gain altitude ponderosa pine, Douglas fir, bear, elk, and deer claim the territory, much as they do in the northern United States.

Altitude and Latitude

Every 1,000 feet of elevation you gain is equal to traveling five degrees of latitude (300 miles) north.

Huey

Bandelier National Monument, New Mexico

Mirocha

Topography

The topography of the area outside the Southwest plays a role in determining both seasonal and daily changes in the weather, which has an effect on the overall climate.

The Sierra Nevada Range in California is the first large topographic barrier incoming Pacific storms encounter. These mountains wring out a good portion of the available moisture. Lying just east of the Sierras—in the lee of the range—are Death Valley and the Mojave Desert. This desert exists because air lifts from the windward side (the side facing the prevailing winds) and then dries out and sinks on the leeward side of the mountains. This is known as a "rain shadow." After crossing the tall Sierra Nevada, this sinking, drying air then moves across the Great Basin and the rest of the Southwest, only leaving additional precipitation when it encounters a mountain range.

The unique topography of the Southwest also affects the weather, creating localized climate niches. For example, areas closer to the bases of the mountain ranges on the windward side generally receive more precipitation. The leeward sides of the mountain ranges have the driest climates.

Mountain locations receive more precipitation throughout the year because they are closer to and therefore receive more of the high and mid-level moisture that moves over them. Some clouds capable of producing precipitation in the mountains often leave adjacent basins completely dry. In the deserts, more frequent precipitation over some isolated mountain ranges helps form unique "islands" of plant and animal life.

At Death Valley National Park, California, the lowest—and often the hottest—point in the western hemisphere, gusts of wind as high as 70 miles per hour push large rocks across clay lakebeds.

Dykinga

Huey

Death Valley National Park's Badwater (282 feet below sea level) lies in the shadow of the Panamint Mountains (11,331 feet above sea level), displaying the stunning change of the Southwest's topography.

Mirocha

Rain Shadows:
Rain shadows are caused by mountains, which force moist air to lift on the windward side. The lifting forces moisture out of the air. Empty of moisture, as the air drops on the leeward side it is much drier.

Temperature

How hot will it get today? That will depend on a number of factors, the most important of which is the sun, followed closely by time of year, cloud cover, and humidity.

The sun does not heat the air directly; rather most of the sun's rays pass through the atmosphere and heat the ground. It is the heat absorbed by the ground that warms the surrounding air near the surface. The clear, dry air and sparse vegetation that typify the desert floor permit the ground to absorb nearly 90 percent of the incoming solar radiation that penetrates our atmosphere, raising temperatures rapidly during the day.

Moister climates generally have more cloud cover and more vegetation; both absorb much of the incoming solar radiation before it can reach the ground.

Clouds can help reduce hot temperatures.

Hot Days

Summer days can get pretty hot in the deserts. Daily air temperatures might be 110° F, while the ground itself might be 150° F or higher. However, escaping the desert heat may be easier than you think.

Heading up into one of the many mountain ranges or onto the mesa tops will afford four to five degrees of cooling for every 1,000 feet of elevation gain. An elevation gain of 4,000 feet can shave as much as 20 degrees off the temperature. For instance, when the thermometer hits 100 degrees on the valley floor in Big Bend National Park, Texas (elevation 1,850 feet) take the 37-mile drive up to Chisos Basin at an elevation of 5,400 feet and you're likely to find temperatures in the low 80s.

Cool Nights

During the day the ground gains heat from the sun. At night the ground loses the stored-up heat by convection radiating it back into space. This is called outgoing infrared terrestrial radiation.

How cool it gets will depend on the humidity, cloud cover, and wind. In a dry atmosphere, overnight temperatures can cool by 40 or 50 degrees from the daytime high, providing welcome relief from the heat. In more humid conditions, humidity and cloud cover prevent the ground from releasing heat into space. Instead, they absorb the heat and send it back toward the earth's surface. This keeps temperatures from cooling significantly. Wind also thwarts nighttime cooling by mixing the air. When the weather is clear and calm all night long, temperatures will cool down more efficiently, making for chilly mornings.

Huey

In the desert, temperatures during the day can be 50 degrees higher than low temperatures overnight.

Faidley

At daybreak near Moab, Utah, Castleton Tower traps the night's cooler air in the valley below to form fog.

Humidity

Humidity is a measure of the amount of water vapor, known as moisture, in the air. Water exists in the atmosphere, even when no clouds are present, in the form of vapor and small invisible droplets. The liquid form is continuously evaporating and reforming. Clouds form when condensation exceeds evaporation and droplets build in size until they become large enough to see.

Relative humidity is a measure of how much moisture is in the air compared to how much it is capable of holding at a given temperature. A relative humidity of 50 percent means that the air—at a given temperature—is only halfway to being completely saturated. At 100 percent relative humidity, the air is completely saturated and is holding as much moisture as it possibly can. At this point water vapor molecules join together to make liquid drops, and clouds form. Near the ground, dew or fog may form.

Dew Point and Relative Humidity

A high dew point temperature indicates high relative humidity, which in turn indicates moist air. A low dew point temperature indicates low relative humidity, which denotes drier air.

When the air temperature is high and the dew point temperature is also high, we say the air feels "muggy" (75 degrees with 75 percent relative humidity would feel muggy).

Heat Index

The Heat Index, developed by R.G. Steadman in 1979, is a measure of how hot it actually feels to us based on different combinations of temperature and relative humidity. The Heat Index is a measure of human comfort (or discomfort) and is useful for planning outdoor activities that might require exercise or prolonged exposure, much like the Wind Chill Index is useful in the winter months in northern climes.

You can see that when relative humidity is low (15 percent or less), as is common in the Southwest, the air will feel several degrees cooler than the actual air temperature.

The Heat Index is based on a person at rest, in the shade, with light winds. Be aware that standing out in the direct sun can make it feel 10 to 15 degrees warmer. Wind can help cool you down, unless the air temperature is above body temperature (98 degrees), then wind will add heat to your body. Exercise or prolonged exposure during a high Heat Index can be dangerous.

Heat Index
Relative Humidity Percentage

Air Temperature	5	10	15	20	25	30	35	40	45	50	55	60	65	70	75	80	85	90	95	100
120	111	116	123	130	139	148														
115	107	111	115	120	127	135	143	151												
110	102	105	108	112	117	123	130	137	143	150										
105	97	100	102	105	106	113	118	123	129	135	142	149								
100	93	95	97	99	101	104	107	110	115	120	126	132	138	144						
95	88	90	91	93	94	96	98	101	104	107	110	114	119	124	130	136				
90	84	85	86	87	88	90	91	93	95	96	98	100	102	106	109	113	117	122		
85	79	80	81	82	83	84	85	86	87	88	89	90	91	93	95	97	99	102	105	108
80	74	75	76	77	77	78	79	79	80	81	81	82	83	85	86	86	87	88	89	91
75	69	70	71	72	72	73	73	74	74	75	75	76	76	77	77	78	78	79	79	80
70	64	65	65	66	66	67	67	68	68	69	69	70	70	70	70	71	71	71	71	72

Dew Point Temperature

The dew point temperature is the temperature to which the air must cool in order for dew to form. Dew (or fog or frost) forms because the temperature has cooled to the point where the air near the ground is completely saturated with as much moisture as it can hold.

The dew point is another measure of the amount of water vapor in the surrounding air. High dew point temperatures (50 to 70 degrees) indicate more humid air. When the weather is constant, the dew point is also fairly constant, and it does not fluctuate during the day and night as air temperature does. When the air temperature equals the dew point temperature, the relative humidity equals 100 percent.

Most of the year, southwestern dew points are low (in the teens and twenties), because the air is so dry. Overnight temperatures rarely cool all the way to the dew point because the nights, especially in summer, are not long enough to achieve sufficiently low temperatures. Thus fog, dew, and frost occur infrequently. However, when dew does form, plants receive some moisture without any rain.

Sunshine and Ultraviolet Rays

Abundant sunshine is what the Southwest is famous for, which is why so many people come to this part of the country for recreation. The southwestern quarter of Arizona receives the highest annual percentage of sunny days in the United States. The area from Yuma to Tucson receives 86 percent of the possible sunshine on an annual basis. That's roughly 314 sunny days per year. Sunny days are at their peak in April and May with 90 to 95 percent of the possible sunshine, producing about 28 cloud-free days in each of those months.

Dykinga

Sandy surfaces, such as those found in Monument Valley, Arizona, reflect up to 60 percent of ultraviolet rays.

Ultraviolet Radiation

The dry air of the Southwest allows more solar radiation to reach the ground. The amount of ultraviolet radiation—the most dangerous part of the total incoming solar radiation spectrum that reaches the earth's surface—depends on the height of the sun in the sky, which is a function of latitude and time of year. The closer to the equator, the higher the amount of UV radiation that reaches the ground. This is because the sun is higher in the sky and the rays are striking the ground more directly.

For every one thousand feet of elevation gain, the amount of UV radiation increases by 2 percent because at higher altitudes there is less atmosphere above you to filter out UV rays.

Ultraviolet radiation at 8,000 feet is 12 percent stronger than the same sunshine you would be getting at 2,000 feet.

(Based on the maximum dosage you could get on a clear day.)

Ultraviolet Index

The Ultraviolet Index was developed by the National Weather Service to give the public an idea of how much solar radiation there could be on any given day. The ratings range from minimal to very high (0 to 10). This index is based on the amount of time it takes a fair-skinned person to burn at noontime if unprotected.

UV Index		Burntime
0 to 2	Minimal	60 minutes
3 to 4	Low	45 minutes
5 to 6	Moderate	30 minutes
7 to 9	High	15 to 25 minutes
10+	Very High	10 minutes or less

Source: National Weather Service

What Affects Your Daily UV Dosage?

Time of Day: The strongest dose of solar or UV radiation is from 11 A.M. to 3 P.M.

Cloud Cover: High, thin clouds allow 50 to 70 percent of the incoming solar radiation through to the ground. Lower, thicker clouds allow only 5 to 25 percent through.

Ground Cover: Sand reflects up to 60 percent of solar radiation back toward the sky. Snow reflects up to 95 percent. Grass reflects up to 30 percent.

Water: At high noon water only reflects back 10 percent of the sun's radiation. Therefore, the water absorbs 90 percent of the sun's rays.

Shade: You are still getting some UV radiation under the trees. As a rule of thumb, if you are casting a shadow, you are taking in UV radiation.

Storm Systems

On an annual basis, the southwestern United States enjoys some of this country's most pleasant weather, thanks to its geographic location.

Worldwide, dry climates tend to be located inland far from large bodies of water. Geographically the Southwest is not that far from the Pacific Ocean. But that part of the ocean, off the coast of southern California and below, near Baja, is usually under the influence of a large, semi-permanent area of high pressure where fair weather is the norm and storm systems do not usually form.

Storms, or low-pressure systems, that affect the western United States normally form farther north in the Pacific Ocean—between forty-five and sixty degrees north latitude—in the Gulf of Alaska.

Prevailing westerly winds, known as "westerlies," carry storm systems from the Pacific across the United States. The Southwest is situated near the southern end of these westerlies, which usually fall between thirty and sixty degrees north latitude as they move from west to east around the globe. Throughout the year, the majority of storms track well to the north of southern California, Arizona, New Mexico, and western Texas.

A winter storm sweeps over southeastern Arizona, fueled by winds and moisture from the Pacific Ocean.

In the winter months, storms traveling with the westerlies will occasionally dip farther south into the deserts of the Southwest, bringing cooler temperatures and snow to the higher elevations.

The deserts of the Great Basin in Nevada and Utah, as well as the Colorado Plateau region, receive the majority of their precipitation in the winter months, though summer thunderstorms account for a small portion as well.

Many locations throughout the Sonoran Desert in southern Arizona and the Chihuahuan Desert in New Mexico and southwest Texas receive the bulk of their annual precipitation from summer thunderstorms. Storms in the winter months (December to early March) usually bring slower, steadier rains, sometimes amounting to less than half of what is received in just July and August.

At Mesa Verde National Park in southern Colorado, winter storms carry snow, which can last through April.

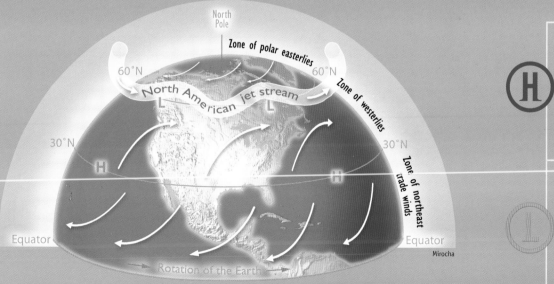

|| Air Circulation
Westerlies carry storm systems from the Pacific across the United States. The jet stream helps guide the westerlies.

High pressure is an area of sinking air, usually devoid of clouds and precipitation, where storms cannot develop.

Low pressure is an area of rising air, which usually generates clouds and storms.

Bean

Mexican poppies and wild hyacinth bloom thanks to
El Niño rains at the Superstition Mountains, Arizona.

Just one year later, La Niña bears much drier conditions.

Bean

El Niño and La Niña

Fluctuations in the way weather
systems track across the Southwest from
year to year can lead to above- or below-
normal precipitation in many locations.
Lately, these fluctuations have been
blamed on El Niño and La Niña.

In brief, El Niño is a situation where
temperatures in the equatorial Pacific
Ocean, off the coast of Peru, are warmer
than normal during December. Hence the
name "El Niño," Spanish for Baby Jesus.
La Niña is when those temperatures are
cooler than normal. These fluctuations in
temperatures disrupt the normal weather
patterns on a global scale.

For the southwestern United States
statistics show that—in general—El Niño
years bring wetter than normal weather,
and La Niña years bring periods of drought.

The Jet Stream

Think of the jet stream as a giant river of the fastest moving air in the atmosphere, at an altitude of about 30,000 feet. This jet stream also serves as a de facto boundary between colder air to the north and warmer air to the south. The bigger the difference in temperature between the cold and warm air, the faster the jet stream. The jet stream also acts to guide storm systems occurring in the lower levels of the atmosphere along its snaking path.

The main jet stream that affects North America spends most of the summer months above Alaska and Canada. In the winter months as days get shorter and temperatures get colder across the lower forty-eight states, this jet moves south, generally hovering over Nevada, Utah, and Colorado. Then the Grand Canyon, the Colorado Plateau, and the Sangre de Cristo Mountains of southern Colorado and New Mexico experience below-freezing temperatures and occasional snowfall.

From year to year the timing and location of this jet stream migration can vary. Much less frequently, this jet stream will dip down into southern California, dragging cold temperatures into the Southwest, sometimes bringing snow to the higher plateau areas of southern Arizona and New Mexico. Even more infrequently, it will dip into the far Southwest and bring a trace of snow to the lower deserts of southern Arizona and New Mexico.

Dykinga

Courthouse Towers, Arches National Park, Utah

Till

The Mojave Desert receives almost all its precipitation in the winter months from a few storms that produce intermittent showers.

19

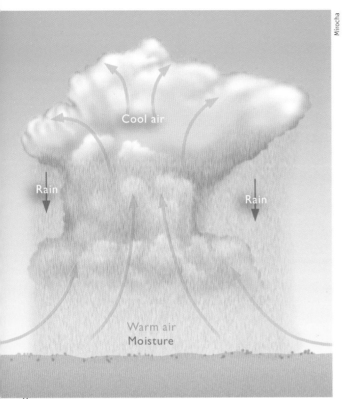

Thunderstorm Formation:
Thunderstorms can form after the sun heats the ground.
Warmer air radiating from the earth then rises into cooler
air aloft. Moisture in the air condenses and falls as rain.

Thunderstorms

"When it rains, it pours," is certainly an accurate description of precipitation patterns in the desert Southwest. And you can never predict where or when it will rain. This is because most of the precipitation comes from thunderstorms.

Thunderstorms have a habit of producing heavy rain in one location, while another location, less than a mile away, will receive little or none. Sometimes a heavy downpour easily produces or even exceeds the average monthly precipitation—and in some cases the annual precipitation!

Thunderstorms require two key ingredients: moisture and instability. First, there must be moisture in the atmosphere to form clouds. Then the atmosphere must be "unstable" enough to keep the clouds developing into a thunderhead.

Thunderstorms start with convective heating. As the ground heats up, warmer air rises into cooler air aloft. If the warm air carries moisture it will condense into droplets and white, puffy cumulus clouds will begin to form. With continued heating, thunderstorms—and precipitation—can develop.

Within developing thunderstorms, the water vapor rises and condenses very rapidly. The rapid condensation causes the cloud to release a lot of energy, which eventually generates gusty winds that facilitate lightning and precipitation.

Prime time for thunderstorms is in the afternoon and early evening hours, just past the hottest part of the day, when the most heat from the ground is rising.

However, thunderstorms are possible at any time of the day or night when a moist, unstable air mass is present, as is common during the summer monsoon.

The earlier cumulus clouds start forming, the earlier thunderstorms are possible. This is an indication that the atmosphere is both moist and very unstable. If you see a very black-based thunderstorm, this shows that more moisture is present, and a stronger, more violent thunderstorm is possible.

Grand Canyon National Park, North Rim, 9:22 A.M.

1:30 P.M.

3:20 P.M.

4:29 P.M.

7:55 P.M.

Cold Fronts

A cold front from a passing storm system can cause a thunderstorm any time of year. One common scenario for cold-front-generated thunderstorms occurs when storm systems moving south and east along the eastern slope of the Rocky Mountains in Colorado, and on across the southern plains of Kansas and Nebraska, force a cold front back into northern New Mexico and Arizona. This is known as a "backdoor" cold front because it sneaks into the area after the main storm has passed well to the north.

While thunderstorms have been recorded almost every month of the year in the Southwest, most occur spring through summer, with a peak in July and August.

Other Thunderstorms

If you are completely baffled by how or why thunderstorms are forming, then perhaps it is just leftover moisture dropped during yesterday's thunderstorms.

After thunderstorms dispense their rain and hail, water on the ground evaporates back into the lower levels of the atmosphere. This "recycled" moisture can lead to the development of new thunderstorms nearby the following day.

Leftover moisture from thunderstorms during the afternoon can also get carried "downstream" in the atmosphere and generate thunderstorms at night or during the next day in an adjacent area that was previously thunderstorm-free.

Thunderstorms frequently occur during the afternoon hours over the Mogollon Rim in north-central Arizona. They drop their rain over the mountains, moistening the atmosphere. Moistened air then descends down over the Sonoran Desert around Phoenix and generates thunderstorms at night when the desert heat is released from the ground.

Raindrops refract light, forming a rainbow over Canyonlands National Park, Utah.

Huey

Virga above southern Arizona mountains

Faidley

Virga is rainfall that evaporates before it reaches the ground. In the Southwest virga looks like a gray veil streaking down from the base of a cloud.

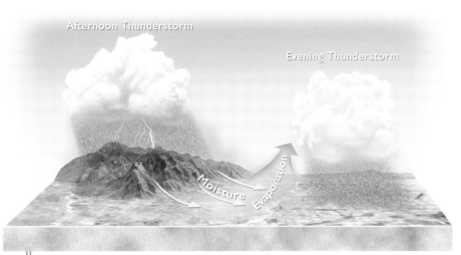

Afternoon Thunderstorm

Evening Thunderstorm

Moisture

Evaporation

Mirocha

Thunderstorms occur five times more frequently over mountains than they do over valleys because mountains serve as barriers to wind flow near the earth's surface. Mountains force air to move upward, providing an extra boost to the air's rising motion.

Lightning and Thunder

Lightning gives the Southwest one of its most dramatic weather shows. Why? Because desert thunderstorms often form above dry air at the earth's surface. So early on in a desert thunderstorm, lightning bolts strike before the air beneath the storm is obscured by rainfall. Even after the rainfall begins, cloud-to-cloud lightning, especially at night, provides a spectacular light show.

As the name implies, thunderstorms also include the sharp report or rumbling sound of thunder, created by the lightning discharge. A crack is usually heard when a bolt strikes the ground. A low rumble is usually associated with flashes occurring within the cloud with no ground strike. Lightning that strikes from a cloud at a great distance from the observer's location may also sound like a low rumble.

Thunder and lightning occur simultaneously, but the difference between the speed of light and the speed of sound presents the flash before the boom. If the strike is right next to you, the flash and boom will be almost simultaneous.

Thunder can be heard as many as 15 miles from the storm. And it is more dangerous than you may think to be outside before a thunderstorm is directly overhead. Why? Because lightning has been known to strike "out of the blue" five to ten miles from the storm.

Cloud-to-cloud lightning travels about 60,000 miles per second above Arches National Park, Utah. About one in four bolts of lightning strikes the ground.

Till

Cordano

Lightning causes the greatest number of wildland fires.

During a five-minute exposure, a photographer captures several lightning strikes in southern Arizona.

Faidley

Faidley

To see how many miles away you are from lightning, count the seconds from the time you see the flash to the time you hear the thunder and divide by five.

An experienced hiker was fatally struck by lightning during a day hike along the highline trail on the Mogollon Rim. His companions noted, "It was overcast, but they were white clouds. We had not seen any lightning, some rain and thunder, but no lightning."

One July, near Tucson, Arizona, a bolt "out of the blue" from a thunderstorm that was estimated to be three miles away fatally struck a golfer.

Hammer-shaped cumulonimbus clouds dump rain on the Grand Canyon.

Monsoon

Desert plants such as the ocotillo, many varieties of cacti, and the creosote bush have all adapted to life with little or no water. Their long, shallow root systems allow them to take advantage of the water from even the briefest shower. When these plants do get a good watering, from a thunderstorm for instance, the ocotillo grow leaves and the creosote bush communicates its familiar odor.

Since it is so dry most of the time in the desert, you might wonder what moisture feeds these watering thunderstorms. Monsoon-related moisture is perhaps the greatest source of fuel for summertime thunderstorms in the southwestern United States.

The term "monsoon" comes from the Arabic word *mausim*, which means "a season." It refers to the large-scale windflow pattern that lasts for an entire season near the Arabian Sea. This is known as the Southwest Asian Monsoon.

The monsoon that affects the United States is sometimes known as the Arizona Monsoon, but it is more properly a northward extension of the Mexican Monsoon. So perhaps the best term used to describe this annual event is the "North American Monsoon."

Our monsoon flow (at low levels) is from the south, coming up out of Mexico. Moist, tropical air—from the southeast Pacific, the Gulf of Mexico, and the Gulf of California—converges over the interior of Mexico in the summer. This more humid, tropical air is then drawn northward over the hot, dry deserts of the southwestern United States and causes frequent, strong thunderstorms, mainly over New Mexico and Arizona.

26

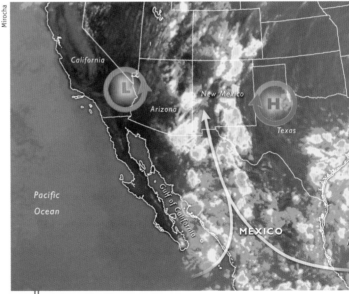

Low and high pressure systems channeling moist air northward into the Southwest

When the dew point temperature averages more than 54 degrees three days in a row, the Weather Service declares the official opening of monsoon season.

Bean

The monsoon flow is brought on when a ridge of high pressure becomes established over southeastern Colorado and the Texas Panhandle. At the same time, a low-pressure system (a thermal low) settles over southern California. The combined circulation from these weather features pumps moisture northward from Mexico into the southwestern United States, bringing on the monsoon.

Often the monsoon will surge into Utah and Colorado. A few times a summer, the monsoon may even push moisture as far north as Montana.

Relative humidity is higher during a monsoon surge, and the air feels more like the southeastern United States.

The "normal" monsoon in the Southwest is July through early September. It begins in southwestern Texas and eastern New Mexico in late June, then extends westward into Arizona by early July in most years.

Huey

Faidley

A flash flood sweeps across southeastern Utah. Floodwaters can cause havoc downstream in areas that haven't received rain.

Flash Floods

It is hard to imagine drowning in the desert. At times it is hard to imagine that it ever rains—but it does, sometimes quite heavily, and often the result is a flash flood.

In the Southwest desert and canyon country, the terrain is rocky and dry. Rain doesn't absorb quickly into the hard-packed earth, especially when it falls at a rapid rate or for a long duration, as it can during a strong thunderstorm. Normally dry canyons, gullies, and washes fill rapidly with the runoff and quickly turn into raging torrents of water.

Flash floods can occur within a matter of minutes or several hours after a rainfall, which has surprised many a person over the years. Statistics show that more people are killed by flash floods in the southwestern United States than by any other weather phenomenon.

It need not be raining where you are for you to witness a flash flood. The runoff from a thunderstorm many miles away may be headed to your location. Rapidly rising water may reach heights of 30 feet or more in confined canyons and is capable of moving large trees and boulders. Just six inches of fast moving water can knock you down. Two feet of water can float a car.

Wind

Many of the storm systems, or low pressure centers, that affect the western United States pass well to the north of the Southwest. The tail ends of these storms and their associated cold fronts move across the Four Corners region, bringing a few clouds and little or no precipitation. In southern Arizona and New Mexico, nary a cloud is seen as these storms pass. The only noticeable change will be an increase in wind.

Why does the wind blow? Well, thanks to high pressure, the Southwest experiences persistently fair weather most of the year. But if a strong low-pressure system passes across the northwestern portion of the country, there is the potential for large pressure gradients to develop.

The difference between high and low pressure determines the strength, or speed, of the wind. The larger the differences in pressure, the higher the wind speed. Consequently, winds are strongest in the Southwest in the springtime, when there is also a larger difference in temperature between the northwestern and southwestern United States.

Typical Windy Weather in the Southwest

A large low-pressure system, a storm, moves across the northwestern United States. At the same time, a large area of high pressure, fair weather, sits over the Southwest. Wind speeds will increase as the low-pressure system moves across the northern and central Rockies. In the Four Corners region, for example, wind speeds of 40 mph or more sweep walls of dust across the landscape.

Faidley

Winds in the Southwest reach near-tornado levels, pushing dust and precipitation across the landscape.

29

Mountain, Valley, Canyon, and Other Winds

In the deserts, a calm morning can turn into a breezy afternoon just because of the sun. The sun's heating of the earth creates a temperature difference between the warmer, less dense air near the ground and the cooler, denser air aloft. This temperature difference produces a corresponding difference in pressure, so that by early afternoon, gusty winds are bending trees and stirring up dust.

The complex topography of the Southwest helps this heating and cooling process, forming complex localized wind patterns. For instance, in the morning hours, air may rise in one area on warm east- or south-facing mountain slopes and sink in adjacent areas over the valley or on the cooler north- or west-facing slopes.

In the evening hours, the reverse may be true, as cool air sinks downslope, and at the same time warmer air rises over the basin bottoms. The side canyons of the Grand Canyon are known for having winds that blow through the night in excess of 30 mph, generated by cool air descending from the rim.

Canyon winds may also be accentuated when there is a large difference in pressure from one area to another, particularly across a mountain range. In many locations throughout the rugged southwestern landscape, winds can howl through canyon corridors, accelerating as the airflow becomes constricted. This is a Venturi effect, much like when water is forced from a bigger pipe through a smaller pipe. You get a faster flow where the water is moving through the constriction.

Dust Storms and Sand Storms

A dust storm is one of the most spectacular weather events you will witness in the desert—from afar. To actually be within the dust storm would be the equivalent of being caught in a blizzard in the Midwest. The very dry desert soil is fodder for these storms, as wind in excess of 35 mph can easily pick up the soil and quickly transport it elsewhere. Sand requires a stronger wind to make it airborne. A real desert sand storm can be a very unpleasant experience, pitting your windshield, sandblasting the paint off your car, and getting into your eyes, ears, and mouth.

Dust storms or sand storms can last from a few hours to 36 hours, and they can stretch for several miles, extending up into the air several thousand feet. Thankfully, there are only two or three per year reported in the southwestern deserts.

Windy weather brings tumbleweeds and pollen.

Wood

Monsoon weather offers more than rain. Thunderstorms are sometimes preceded by great walls of dust.

In Monument Valley, a windstorm obscures the highway.

Agriculture and development have increased the severity of dust storms in the Southwest.

Faidley

On sunny days dust devils seem to develop out of nowhere. Temperatures need only be above 80 degrees.

Dust Devils

Unlike a tornado, which is associated with a severe thunderstorm, dust devils develop on sunny days in the absence of clouds. You will see them spinning across the desert floor, gathering a whirlwind of dust or debris, often with a distinct funnel shape. Temperatures need only be above 80 degrees for dust devils to form. Picture a rising column of warm air going about its merry way across the desert, when suddenly it bumps into a bush or some irregularity on the surface. This initiates a change in the horizontal wind direction, which sets the air into a spin. As Ben Herman of the University of Arizona Atmospheric Sciences Department describes, "There is always some angular momentum present in the air as it moves along (just like a spinning figure skater), which initiates the spinning motion of the dust devil."

While the average dust devil is less than 3 feet in diameter, with winds of 45 mph, and lasts for less than a minute, researchers have reported dust devils as large as 100 yards wide swirling across southwestern flatlands at up to 75 mph for as many as 20 minutes.

Near Prescott, Arizona in June 1953, two boys were struck by a dust devil. One sustained a black eye, the other fractured two vertebrae in his back.